HOW TO USE COOPERATIVE LEARNING IN THE MATHEMATICS CLASS

HOW TO USE COOPERATIVE LEARNING IN THE MATHEMATICS CLASS

by
Alice F. Artzt
and
Claire M. Newman

Queens College
of the
City University of New York

NATIONAL COUNCIL OF TEACHERS OF
MATHEMATICS

Library of Congress Cataloging-in-Publication Data:

Artzt, Alice F.
    How to use cooperative learning in the mathematics class / by
Alice F. Artzt and Claire M. Newman.
        p.   cm.
    Includes bibliographical references.
    ISBN 0-87353-293-7
    1. Group work in education.   2. Mathematics—Study and teaching.
I. Newman, Claire M.   II. Title.
QA16.A78   1990
510'.71—dc20   90-33599
                                                                CIP

Printed in the United States of America

CONTENTS

Introduction 1

    Cooperation in the Classroom 1

    Cooperation in the Mathematics Classroom 1

What Is Cooperative Learning? 2

    Cooperative Learning Schemes 3

    Research Results: Positive Outcomes 4

Incorporating Cooperative Learning in the Mathematics Class 5

    Getting Started: An Example 5

        Discussion 8

    Other Opportunities for Cooperative Learning in the Mathematics Class 9

        Review of Homework 9

        The Developmental Lesson 9

        The Review Lesson 11

        Test Review 12

        Enrichment 12

Making It Work 14

    Group Formation 14

        Assignment of Students 14

        Duration of Group 15

        Size of Group 15

    Features of Successful Groups 15

        Mutual Dependence 15

        Verbal Interaction 16

        Interpersonal and Group Skills 17

        Individual Accountability 18

        Incentives and Rewards 18

    The Teacher's Role 19

Concluding Remarks 20

Sample Activities 21

Bibliography 72

INTRODUCTION

Cooperation is critical to the progress of human civilization. It is as relevant to the interactions of nations as it is to the operations of business organizations. It is as important to the success of sports teams as it is to the advancement of science. It is as valuable to the functioning of families as it is to the relationships of individuals. In contemporary society, cooperative individuals are more likely to attain their personal goals. Moreover, the diversity of our population requires the individual to accept and work with others who are different. Regardless of race, ability, or handicap, all students must learn that it is in their interest to cooperate for the common good.

Cooperation in the Classroom

For individuals to cooperate they must develop certain skills and understandings. Students who pass through a standard school program do not necessarily develop these skills. Unless schools engage in specific activities that teach cooperation, students may lack the skills they need to function in an adult society. The classroom is a natural setting for cooperative learning activities. Students who have the opportunity to work in small groups can begin to practice the cooperative skills necessary for the group members to solve problems together. Furthermore, each member of the group can learn the content of the curriculum through his or her interactions with the other members of the group.

Cooperation in the Mathematics Classroom

Support for this kind of cooperative learning experience comes from the National Council of Teachers of Mathematics (1989). *Curriculum and Evaluation Standards for School Mathematics*, adopted by the Council, recommends that teachers provide opportunities for students to work together in small groups to solve problems. In this way students can talk about the problem under consideration, discuss solution strategies, relate the problem to others that have been solved before, resolve difficulties, and think about the entire problem-solving process. According to the *Standards*, "Small groups provide a forum in which students ask questions, discuss ideas, make mistakes, learn to listen to others' ideas, offer

1

constructive criticism, and summarize their discoveries in writing'' (p. 79). Group assignments enable learners to work together, helping one another integrate new knowledge with prior knowledge and discover their own meanings as they explore, discuss, explain, relate, and question new ideas that arise in the group.

Just what is cooperative learning, and how can cooperative learning methods be incorporated into the mathematics class?

WHAT IS COOPERATIVE LEARNING?

Cooperative learning involves a small group of learners, who work together as a team to solve a problem, complete a task, or accomplish a common goal. There are many different cooperative learning techniques; however, all of them have certain elements in common. These elements are the ingredients necessary to insure that when students *do* work in groups, they work cooperatively. First, the members of a group must perceive that they are part of a team and that they all have a common goal. Second, group members must realize that the problem they are to solve is a group problem and that the success or failure of the group will be shared by all of the members of the group. Third, to accomplish the

group's goal, all students must talk with one another—to engage in discussion of all problems. Finally, it must be clear to all that each member's individual work has a direct effect on the group's success. Teamwork is of utmost importance.

It is *not* sufficient to direct a group of students to separate into small groups and work on a problem or a set of problems. It is *not* cooperative learning if students sit together in groups and work on problems individually. It is *not* cooperative learning if students sit together in groups and let one person do all of the work. True cooperative learning requires the guidance of a teacher who can help students understand group dynamics, develop the cooperative learning skills they need, and learn mathematics by working together in groups. Cooperative learning capitalizes on the presence of student peers, encourages student-to-student interaction, and establishes symbiotic relationships among team members.

Cooperative Learning Schemes

The cooperative learning literature sets forth many cooperative learning structures that researchers have developed and studied.

Slavin developed and studied several methods he calls Student Team Learning (1980). Using one of those methods, Student Teams–Achievement Divisions (STAD), the teacher presents a lesson, and then the students meet in teams of four or five members to complete a set of worksheets on the lesson. Each student then takes a quiz on the material, and the scores the students contribute to their teams are based on the degree to which they have improved over their individual past averages. The highest scoring teams are recognized in a weekly class newsletter. Another method, Teams-Games-Tournament (TGT), is similar to STAD, but instead of taking quizzes the students play academic games as representatives of their teams. They compete with other students having similar achievement levels.

In Jigsaw II, a modification of an earlier cooperative learning method by Aronson (1978), each team member is assigned a special topic to learn. The student meets with the team members of the other teams who are learning the same topic. After exchanging ideas and information the student returns to his or her own team to teach teammates what he or she has learned. The students take a quiz on the material, and their scores are used to form individual and team scores.

Johnson and Johnson support a cooperative learning technique they have named Learning Together (1975). In this strategy, students meet in heterogeneous groups of four or five members and work on assignment sheets. When the group agrees on solutions to the problems assigned, the members submit a single answer sheet for the entire group. Recognition is based on the group product. However, Johnson and Johnson make it clear that the following conditions must exist for group learning to take place: positive interdependence, face-to-face promotive interaction, individual accountability, social skills, and group processing. Much of the discussion in this book is consistent with the Johnsons' work.

Research Results: Positive Outcomes

The positive outcomes of cooperative learning strategies have been well documented by studies conducted at all grade levels and in all subject areas. Some of these research projects have compared cooperative learning methods with whole-class learning methods. Others have examined the individual and group processes that occur within cooperating teams. Most of the accepted cooperative learning strategies promote the formation of groups that are heterogeneous in a multitude of ways; that is, groups may consist of students of different abilities and ethnic backgrounds, and they may include students who are handicapped. Such heterogeneity leads to positive academic and social outcomes.

There is strong evidence to indicate that cooperative learning is beneficial for students across many dimensions. Cooperative learning capitalizes on the powerful influence of peer relationships. By promoting interaction within the group, cooperative learning teaches students to be supportive and accepting of students who are different. When placed in a working group, students of different abilities, cultural backgrounds, and physical makeups have a common ground for discourse. The process of working together and getting to know one another has proved highly successful in removing artificial barriers and prejudices that ignorance and unfamiliarity create.

It is well documented that positive attitudes toward mathematics play a major role in a student's ability to learn mathematics. Research indicates that cooperative learning experiences in the mathematics classroom foster improved attitudes toward the subject matter and toward the instructional experience. The individual builds confidence in his or her own ability to do mathematics, thereby relieving the math anxiety that many students experience. The collaboration that takes place in a cooperative group gives each student the opportunity to provide help and receive help in a private, nonthreatening way. The small-group setting provides a comfortable social environment.

When students of different abilities are grouped together, more frequent giving and receiving of explanations take place. Contrary to what one might expect, the high-ability student derives as much benefit from group interaction as do the low- and average-ability students. The verbal communication of mathematics is a means for students to become actively involved in learning mathematics. To give a mathematical explanation to one's peers, a student must understand the material with far more depth than is required merely to produce an answer on a worksheet.

The importance of peer relationships within a classroom should not be taken lightly. Children of all ages are influenced by their peers. If the classroom is structured cooperatively, the peer influence can be used to positive ends. Students want their peers to do well. After all, an individual's success depends on the success of the group. Students want their peers to be prepared with the work and to be attentive and productive in class. Peer pressure for academic achievement is one of the most important factors that contribute to the many positive learning outcomes of cooperative learning. Students are motivated to do well, to be prepared with their work, and to be attentive during classtime, for these are the behaviors that lead to peer approval and group success.

Cooperative learning strategies have been credited with the promotion of critical thinking, higher-level thinking, and improved problem-solving ability of students. Current research examining behaviors that occur during group problem-solving sessions seems to indicate that groups engage in behaviors similar to those exhibited by mathematicians when they solve problems; that is, groups monitor their own thoughts, the thoughts of their teammates, and the status of the problem-solving process. They thereby often avoid the thoughtless wild-goose chases so characteristic of novice problem solvers working alone.

There appear to be good reasons to incorporate cooperative learning in the mathematics class, and there are many techniques and structures that others have found useful. Now it is appropriate to explore just *how* to get started.

INCORPORATING COOPERATIVE LEARNING IN THE MATHEMATICS CLASS

Getting Started: An Example

There are many ways of introducing cooperative learning in the mathematics lesson. The teacher has many questions to consider before getting started. At what point in the lesson should the students form groups? Should the groups be used for review of the homework? Review of the classwork? Experimentation leading to the development of the lesson? Test review? How should the groups be formed? Should the students be allowed to choose their own groups? Should the students be assigned to groups randomly, or should they be assigned using specific criteria?

Once the groups are formed, the teacher has other questions to consider. How should the class be structured to provide maximum participation of students within their groups? Should the groups compete with one another? Should there be rewards or incentives? What should be the criteria for receiving an award?

None of these questions has one correct answer. There are many options, and this book will discuss some of the possibilities the teacher should consider.

One way of getting into the cooperative learning mode is to try it out with a homework assignment. Try this: Advise students that they will be engaging in group problem solving. Suggest that they form their own groups of three to five members. (The teacher's role in facilitating the grouping will be discussed later.) To begin, have the members of each group compare their solutions to the previous night's assignment. Ask the students to discuss their work with other members of the group and to come to an agreement on the best solutions. Then ask each group to submit one set of solutions. Next, lead a discussion based on the difficulties the students have encountered. Keep a record of the performance of each of the groups and make the standings available to the class. To add enjoyment to the new class structure, ask the members of each group to devise a whimsical name for their group. What follows is one possible scenario.

Three days with Ms. Jones

Time: The beginning of a math lesson, several weeks into the beginning of the year.

Day 1

1. Ms. Jones says: "We always go over the homework in the same old way. Let's try something new! Why don't you all take out your work and arrange yourselves in groups of about four each?"

2. After the students are sitting in groups, Ms. Jones says: "Now all of you examine your homework and see if you can come up with a set of solutions that you, as a group, believe are correct. When you have finished, hand in the solutions you have agreed on, and we will discuss your work."

3. When all of the group papers have been submitted, Ms. Jones leads a discussion about the solutions the students have been unable to agree on.

4. Ms. Jones says, "Tonight I will read your group papers, and tomorrow I will tell you how well your group did. I will need a way of referring

to each group; so before tomorrow please have a group meeting and agree on a name for your group. I hope you had fun doing this because I know I really enjoyed hearing you discuss mathematics together. Let's do this again tomorrow.

Day 2

1. Each group submits its group name and a list of group members.
2. Ms. Jones returns the group papers from the previous day and announces how well each of the groups did on the work submitted. She leads a discussion about those problems that she or the students feel need additional clarification.
3. Students are asked to form their groups and to agree on the homework due for that day.

4. Ms. Jones walks around the room checking that the students are prepared with their work, are participating in the group's discussions, and are helping one another understand the work.

5. Each group submits one set of solutions.

6. Ms. Jones discusses problems requested by students, or those some groups found difficult.

7. The class members spend a few minutes discussing their reactions to working in groups.

8. Ms. Jones proceeds with the new lesson.

Day 3

1. Ms. Jones displays a chart with the team names and members.

2. Ms. Jones explains the scoring procedure. A tally system is used. The groups are ranked for the day on the basis of the number of correct solutions they had on the assignment. The group with the most correct solutions is ranked first and receives one point; the team with the second highest number of correct solutions is ranked second and receives two points. Tying teams receive the same number of points.

3. Ms. Jones returns the previous day's group work, announcing the group's name, the group's place standing on that assignment, and the new cumulative score from the past two days. At the end of the week the group with the *lowest* cumulative score is considered the most successful group.

Discussion

The scenario given above demonstrates one way that groups can be used to review homework. In a brief three-day period Ms. Jones has given her students the opportunity to work in small groups in an environment that has valued cooperation as a means of success. When students have the opportunity to check their homework within the privacy of the group, trivial difficulties can be ironed out within the group. The whole class does not have to direct its attention to these issues. While the students participate in their groups, the teacher is free to provide individual attention where it is needed. The teacher may check a student's work, ask a key question, pinpoint some of the difficulties a student is having, or make suggestions that will help a student develop some of the behaviors needed to work effectively with others.

Having the groups discuss their homework, work that each student has done individually, maximizes the probability that each student will have something to contribute to the group. Students who work together are dependent on one another in a very positive sense. Each student depends on all group members to do the homework assignment and to do a good job of it. Knowing that they must achieve as a group causes students to be supportive of the academic achievement of their peers.

The members of Ms. Jones's class have had the opportunity to discuss mathematics with one another in a small-group setting. Students have worked together in a pleasant, nonthreatening environment, and they have all benefited from the experience. Students have learned mathematics *and* have picked up some of the skills needed to work cooperatively. They have begun to see how cooperative learning enhances their own individual learning.

It is not necessary to discuss all of the assigned work. Often it is only the most challenging homework that need be discussed with the whole class. At other times the teacher may wish to single out those exercises that are most suitable for group or class discussion because they illustrate a particular idea or method. Other solutions can be made available or discussed at student request.

Other Opportunities for Cooperative Learning in the Mathematics Class

Review of Homework (Another Approach)

There are other ways that groups can be used to review homework. For example, the teacher may wish to have particular problems or applications demonstrated at the board. Each group can be held responsible for discussing one solution that they have agreed on. One member of the group can write the solution on the board while another student, the spokesperson for the group, explains the group's work to the class. Since the teacher chooses the spokesperson and group members do not know who that will be, each student must be prepared to be that spokesperson. Other groups may challenge the work and provide constructive criticism. Any group member may respond to such criticism by accepting it or defending the group's original solution.

It is not unusual for certain groups to be more successful than others. To maintain the enjoyment and excitement of group work, the more challenging problems may be assigned to the more successful groups. However, it is not just the successful groups that enjoy the challenge and prestige of more difficult problems. All groups enjoy work that is both challenging and at their level of ability.

The Developmental Lesson

Cooperative learning is a versatile approach that can be used at many different points in the lesson for many different purposes. In a developmental lesson some new concept, technique, or generalization evolves during the course of the lesson. In the direct instruction format the teacher usually develops, explains, or demonstrates a new technique that can be used to compute, solve equations, draw graphs, prove theorems, and so on. The teacher then asks the class to apply their new knowledge by trying several similar applications on their own. This gives the students the chance to see how well they understand the new material. The students are expected to ask questions about the work so that the teacher can clear up any misunderstandings they may have. Unfortunately, students are reluctant to ask questions in front of the whole class. They fear the embarrassment of being wrong or appearing stupid.

The practice work that is assigned after the direct presentation of a lesson lends itself to group work. In this approach, the teacher assigns the applications in the usual manner, allowing the students time to work on them individually. Instead of being called together for a whole class review of what they have done, the students meet in groups to discuss and agree on the assigned work. Each group is responsible for handing in one copy of the solutions agreed on. After the work has been submitted, the teacher leads a discussion of those applications needing clarification. Having discussed them in their groups, the students are eager to clear up any misunderstandings they may have.

As an added assignment the group may be asked to respond to two questions: "What have we learned today that we didn't know before?" and "What would we like to know as a result of today's work?" The two suggested questions provide each group with an opportunity to summarize the lesson and provide the teacher with material for future lessons. Educators have been discovering that there are advantages to having students write in all subject areas. Asking the groups to write a sentence or two about what they have learned on a particular day gives the students the opportunity to reflect on the work they have done. They are able to integrate new ideas into their previous knowledge base and see how each new idea fits into the whole mathematical picture. It may also suggest what is still to come, what it is they still don't know. Students who participate in stimulating small-group discussions begin to assume some responsibility for their own learning.

A developmental lesson designed for guided discovery leading to mathematical generalizations is well suited to the cooperative learning mode. Each member of a group can be assigned a different task. After completing their different assignments, the group members record their results on a group record sheet and look for a pattern. The group is asked to construct a general statement that describes their results. For example, a group may be given a set of triangles, in various shapes and sizes. They are to measure the angles of each triangle. Each student is given a protractor and several different triangles to measure. There are right triangles, obtuse triangles, isosceles triangles, equilateral triangles, and scalene triangles. Some triangles are small and some are large. After each student has recorded the measurements of his or her triangles, the group meets to record and discuss the results. They try to discern a pattern in the data they have collected and recorded. They agree on a general statement about triangles—all triangles. Once these statements have been written on a transparency (for use on the overhead projector) or on the chalkboard, the statements are compared and discussed by the class. The teacher guides the class until they agree on the statement of a generalization that is the culmination of the day's lesson.

This approach has a number of advantages. Since each student is responsible to the group to do his or her own piece of the work, all group members will be actively involved in the group's work. Dividing up the work this way saves valuable class time; the group obtains more data in less time. The group is using inductive reasoning and practicing behaviors that enhance their problem-solving skills. They are learning, also, that cooperation pays off.

The Review Lesson

The small group setting is particularly useful for review and reinforcement. The Team Mathematics Bee is a variation on the classic spelling bee. Each group is a team competing for the class championship. Teams are numbered, 1, 2, 3, . . . The teacher presents the same problem to each team. The members of the team must agree on the solution to the problem. Team 1 is asked to respond to the problem. A spokesperson for the group is chosen by the teacher to explain the team's solution. If the solution is correct, the team gets one point. If the solution is *not* correct, Team 2 is asked to offer its solution. Each team, in order, is asked to present its solution until a team presents the correct solution. That team receives one point for the solution to that problem. The teacher then presents another problem for the groups to consider. The next team (after the one that solved the last problem) gets the first shot at responding to the next problem. The process continues until the groups have worked on the review problems the teacher has chosen to present. Time should be allowed for discussion of questions that arise and ideas that need clarification. At the end of the game, the team with the most points is declared the champion group. However, it is not only the champion group that has won. All students will have put their effort into an enjoyable review lesson, and each student will have benefited.

This game will be most effective when it is structured carefully. When students are given a problem that they must work on in a group, the potential exists for some students to sit back and let others do the work. This is less likely to happen if each student has had the opportunity to work on the problem before the group discusses it. It is important for students to realize that every team member must be able to discuss the team's solution, since the teacher may choose any one of them to be spokesperson for the team. The prospect of having to explain the solution to the class means that students must help one another understand each problem and work together on solving each problem. Cooperation becomes an important method for achieving group success.

A variation of the Team Mathematics Bee is the Challenge Mathematics Bee. In this game, the teacher does not determine whether or not a solution is correct. Instead, one team presents its solution. The next team, in order, may challenge the solution. If a team challenges a solution and is correct, that team receives two points. However, if a team challenges a correct solution, the team with the correct solution gets two points. Once again, the spokesperson for each team is selected by the teacher. If the second team does not wish to challenge, the opportunity to challenge passes on to each subsequent team. If there are no challenges, the first team receives one point. In this game students are given the responsibility for their own learning. It is the students, not the teacher, who decide which solutions are correct. Student attention level is high, and students soon learn that success depends on cooperation within the group.

Test Review

A sample test can be assigned for homework. The students then meet in groups to discuss the test and to deepen their understanding of the concepts and techniques that will be tested. By working on the sample test individually, each student comes to the group discussion with an accurate picture of his or her understanding. Students are able to prepare themselves and other members of the group for the forthcoming test. Each student is glad to participate in the review so that he or she will do well on the test. Once again, each group agrees on the solutions to the problems and hands in one group paper. The teacher allows time for the whole class to discuss those areas that need clarification.

Enrichment

Enrichment activities have an important role in the mathematics program. When enrichment is an integral part of their work in mathematics, students of all ages and abilities learn to appreciate mathematics as a living, useful, interesting subject.

Students want to know how mathematics is used in various careers, in the world around them, in the stock market, in the supermarket, or in the daily newspaper. They are proud to learn that their ancestors (men, women, Hispanics, blacks, etc.) have made important contributions to the development of mathematics and that people of their own background are making contributions right now. They are interested in seeing how mathematics is involved in music, art, and the sciences. For many students intriguing problems or puzzles not only offer exciting sources of recreation but also spark their interest in mathematics.

Group work is an excellent way of incorporating enrichment experiences in the mathematics class. To spark student interest in a new topic, cooperative learning groups can investigate the historical development of the topic. The members of the group should divide up the work. One student looks up the dateline in the development of the topic. Another student is responsible for uncovering the mathematicians who were instrumental in the development of the topic. The group will want to have a person look for anecdotes and events related to the topic. Finally, it will be of interest for a group member to investigate how knowledge of this topic has affected the world as it is today. This group project might culminate in a bulletin board display or a report (written or oral).

In a class studying statistics, each group can conduct a survey that answers a question or deals with an issue of interest to students in the class or school. Each group collects data and records them in tables. They study the data and display them in graph form. Finally, each group interprets its data and reports its finding to the whole class. There is much to be done, and the members of the group must cooperate to accomplish the task. Individuals will be working on their own some of the time, for there will have to be a division of labor. When all the pieces come together, however, it will be the group that gets credit for the final product.

Cooperative learning groups can engage in recreational mathematics unrelated to current class work. Such activities can take the form of puzzles, games, or problems that challenge the students to do creative problem solving. This work can be organized in many different ways. On Monday the teacher posts one problem-solving activity on the bulletin board. The groups are asked to solve the problem by the end of the week. At the end of the week those groups that claim to have solved the problem present their solutions. The teacher chooses a spokesperson for a group. The spokesperson must present the group's solution satisfactorily before the group can get credit for having solved the problem. The teacher can also have a grab bag of problems; each group selects a different activity to work on. Again, the group gets credit for a solution only if the spokesperson selected by the teacher is able to explain the work with clarity and understanding.

Group problem solving has many advantages. Members of the group engage in brainstorming, an activity that enables all members of the group to participate in a free flow of ideas. The cooperative learning attitude provides a secure environment for everyone to make a contribution. The student who is poor at solving problems has the opportunity to engage in the problem-solving process along with peers who are more able. Not only do all students learn how to solve problems, but they also share in the excitement the group experiences when the problem has been solved.

MAKING IT WORK

Group Formation

To maximize the benefits of cooperative learning groups, the membership should be heterogeneous in ability and personal characteristics. The group must stay together long enough for cohesiveness to develop. A successful group will be small enough for everyone to be needed but large enough to permit a diversity of ideas and skills.

Assignment of Students

The most effective way of ensuring heterogeneity is for the teachers to set up the groups. Teachers know their own students best and can see to it that they place readers with nonreaders, task-oriented students with non-task-oriented students, high-ability students with medium- and low-ability students, minority students with majority students, non-English-speaking students with those who speak English, handicapped students with nonhandicapped students, and females with males. It would be desirable to ask students to indicate which peers they would like to work with and consider their wishes when groups are being formed. It is important that students be happy in their groups if they are to work well.

There are various other grouping strategies one can use. In the example given earlier, Ms. Jones simply asked her students to form groups. In this self-selection method, students usually select friends or peers who are very much like themselves—same sex, same ethnic background, same ability, or the like. This tends to result in homogeneous groups, and, very often, certain students are left out entirely. Since cooperative learning works best when groups are heterogeneous, self-selection is not a good strategy unless the teacher sets up certain restrictions that result in heterogeneous groups.

Students can be randomly assigned to groups. This method is particularly useful at the beginning of the school year when the teacher has little information about the class. Students can count off, or names can be placed on slips of paper and drawn from a bag. Older students can learn about the use of a table of random numbers while the teacher uses the table to assign them to their groups.

Duration of Group

One of the criteria for group success is the durability of the group. It takes time for group cohesiveness to develop. When students know that their group will be together for some time, they realize that they must improve their interpersonal skills so that they can function effectively. Over a period of time they come to recognize their responsibilities to the group. This is often facilitated by the peer pressure that group members exert on one another.

Cooperative learning groups may stay together during a unit of work, a semester, or a year. Certainly there is something to be said for giving students the opportunity to use their group experiences in a new group setting. A new group means new relationships, new ideas and opinions, and new friends. Once a teacher has had some experience in incorporating group work into the class, she or he will know when it is desirable or appropriate to form new groups.

Size of Group

The size of a group affects its ability to be productive. Experience has shown that groups of three to five students work well. Although it is often advantageous for students to work in pairs, a cooperative learning group consisting of just two members is at a distinct disadvantage. Interaction is limited, and the group is vulnerable to the absence of either of its members.

Conversely, a cooperative learning group should not be too large. If a group has too many students, it becomes very difficult for the group to function effectively. The most vocal students tend to take over, and the quiet ones recede into the background. In a large group it is difficult for everybody to air their ideas. Furthermore, it is difficult for a large group to get organized, to coordinate the work of its members, and to reach agreement.

Features of Successful Groups

For cooperative learning to succeed, students must function effectively in the group. Members of successful cooperative groups learn to use those skills that are important to group success.

Mutual Dependence

For cooperative learning to succeed, the students in a group must perceive themselves as dependent on one another. To be successful in achieving the group's goal or completing the group's task, each member of a group expects each of the others to make a contribution. It is not enough for one person to be generous or altruistic while others accept what is given. Cooperation is based on reciprocity. Maintaining effective working relationships among group members requires each student to appreciate the value of reciprocation. Each student must be prepared to give as well as to receive.

A cooperative learning group functions well when students are concerned not only about themselves but also about the other members of the group. Students engage in peer teaching because they acknowledge that each member of the group must understand the material. Each student recognizes that group members expect him or her to complete the assigned work and to make a contribution to the group. Students help one another. One student explains a difficult concept to another in his or her own words. Group members share resources and act as resources for one another. Students divide up their responsibilities in the group. Students praise one another for their work. Students encourage one another to participate. Even those who are usually silent are made to feel that the group relies on them to participate in the group's activities. It's "all for one and one for all" because that is what makes group success possible.

Verbal Interaction

For cooperative learning to succeed, students must talk to one another. There must be a regular exchange of ideas. In a successful cooperative learning group, students communicate, explain, and justify ideas and, when necessary, engage in intellectual conflict. A group should be involved in verbal conflict—conflict over ideas. Group members should be critical of ideas but not people. Students should feel comfortable about disagreeing openly because they know that such controversy in no way damages their standing in the group. They know, too, that controversy strengthens their own understanding and helps the group reach consensus.

When a group is functioning well, each member has something to say and the opportunity to say it. Each person listens carefully to what other group members are saying and tries to understand ideas with which he or she disagrees. Each group member tries to bring out all ideas before accepting a strategy or an answer. Before deciding on a problem-solving strategy, students brainstorm ideas and discuss the approaches that seem appropriate.

Students in a successful cooperative learning group are eager to check with one another to be sure that each person in the group understands the material, agrees with the results or conclusions, and is able to represent the group as a spokesperson. Students ask one another for help or clarification. Students also offer to explain or clarify mathematics concepts and methods. Material is summarized aloud. Students explain to one another how new knowledge relates to the material that has come before. Students ask questions, and students answer questions.

Learning in a group necessitates verbal interaction. To work together, people must talk with one another about the task at hand and about what they are doing and thinking. The quality of the verbal interaction is an important factor in the success of the group.

Interpersonal and Group Skills

The quality of certain interpersonal and group skills determines how well people work together. Students need to develop cooperation skills if they are to function well in groups, be they cooperative learning groups, families, business organizations, or communities. For cooperative learning to succeed, students must master those collaborative skills that enable them to work effectively with others, no matter what their abilities or personal characteristics.

Communication is necessary for people to cooperate. People must be able to send messages so that others will understand their ideas and feelings. These messages can be verbal or nonverbal, but they must be conveyed in such a manner that the person receiving the message has a clear understanding of what is meant. The receiver should be certain that he or she understands what is being communicated. The receiver does this by restating what has been said or by asking one or more questions designed to elicit a better understanding of the message.

Trust is another important ingredient of cooperation. When there is trust, people are willing to be open in the communication of their ideas and feelings. A person expresses trust when he or she is willing to share ideas with others in the group. People trust other members of the group when they perceive that others are accepting of their ideas and are warm toward them. They feel that the value of their contributions is recognized by others. Trust is nourished when people support one another and convey the belief that each person is capable of being a productive member of the group. The person who praises others for their contributions, encourages everyone to participate, shares materials with others, and offers to help others is a person who can be trusted.

For members of a group to cooperate and to complete their task, group members must share the responsibility for taking on a leadership role. Someone must help the group get started and stay on task; someone must coordinate the group's efforts; someone must give direction to the group's work; and someone must encourage participation. A person who can accomplish any of these goals in a warm, friendly, nonthreatening manner displays the skills of a leader.

It is not unusual for differences and disagreements to arise even though the group is working cooperatively. Group members need the skills to manage such controversies. They must ask questions. They need to clarify differences. Each person must be patient and exert self-control. Once all ideas have been discussed, group members must be willing to compromise—to integrate different perspectives into a single group solution that is acceptable to all. Such skills of conflict management are essential to the functioning of any group.

Individual Accountability

For cooperative learning to succeed, there must be individual accountability. In cooperative learning groups, responsibility for each person's learning is shared by other group members. Group members are expected to provide help and encouragement to one another. The emphasis is on working together and learning together. Nevertheless, individual students are held accountable for their own learning and for their contributions to the group. Each member of the group is responsible for mastering the material. Written and oral tests assess each student's progress and learning. Each student is expected to bring assigned work to the group so that the group can complete its task. The group holds its members individually accountable for such assignments. Each student is responsible for presenting a report of the group's work to the class. Individuals may not sit back and defer to others in the group. They are expected to learn and to participate in the group's work.

Incentives and Rewards

What is it that makes group learning attractive to students? There are many reasons for students to want to work and learn together in groups. Most important are the incentives and rewards that provide the intrinsic motivation for cooperative learning. The social aspects of group work are enjoyable. Students form new friendships and learn to appreciate differences—differences in ability, differences in personal characteristics, and differences of opinion. Students find that learning together is fun and that being part of a group is exciting. The student who helps others experiences gratification in giving. The students who know that they can depend on other group members for help and support are relieved of the anxiety often experienced by those who don't understand the work. There is a real sense of satisfaction in learning, achieving, and solving problems together. Cooperative learning can be truly rewarding.

As an additional incentive you may wish to give students group rewards. These rewards need not be expensive or elaborate. The teacher's praise and a smile may be all that are needed. A smiling-face sticker may be even better. Gold and silver stars have worked well in the elementary grades for a long, long time. Lists of rewards can be brainstormed by students. A handmade certificate can be awarded to the Group Champ of the Week. Applause should follow the presentation of the award and the group's name should be placed on the bulletin board for all to see.

The Teacher's Role

The teacher plays a vital role in the implementation of effective cooperative learning. It has already been pointed out that the teacher is responsible for the formation of groups and for the ways that incentives and rewards are used. Implicit in the assignment of a group problem is the teacher's responsibility to explain the assignment, the academic expectations for the group, the expected collaborative behaviors, the procedures to follow, and the definition of group success.

The materials and the instructions for their use should be structured in such a way that each student in the group does something to contribute to the group's work. The problem should be set up so that group members are dependent on one another to complete the task. Requiring a single group product, for example, emphasizes the need for cooperation. It should be made clear that group members are responsible for one another. Each person is expected to learn the material and to help others learn the material.

The teacher will have to devise ways to hold each student accountable for learning the material. This can be done by giving frequent quizzes, randomly selecting a group spokesperson to explain the group's solution, and randomly picking papers to grade.

The teacher, as class manager, must see to it that the room is organized in such a way that the members of a group are close enough to one another to work together comfortably and talk with one another quietly. The groups must be separated so that they don't interfere with one another.

Teachers must monitor the groups while they are in progress and provide assistance as it is needed. When a group is functioning poorly, the teacher will want to intervene to help students with the cooperative skills they need. Once these skills have been identified and discussed, the teacher will want to see how well the group is practicing them and whether it is functioning more effectively. The teacher should provide feedback so that students know how well they are doing. The teacher may ask the groups to monitor their own performance by answering questions about the group's behavior and functioning. Is each person participating? Are students helping one another? Are they handling conflicts well?

Needless to say, as teachers become comfortable with the cooperative learning approach, they will decide themselves how best to facilitate the cooperative learning process.

CONCLUDING REMARKS

Cooperative learning has much to offer for the mathematics class. Students enjoy discussing mathematics with other students; they benefit from their interaction with their peers as well as with the teacher.

In the traditional teacher-centered classroom, depicted in figure A, the teacher is clearly at the center of all class activity. All lines of communication are between individual students and the teacher. In contrast, figure B depicts the cooperative learning classroom. This shows that although the teacher still retains a central role, he or she is no longer responsible for all mathematical discussion. The students help the teacher meet the demands of dealing with a whole class by serving as resources for one another within their groups.

Cooperative learning is an instructional strategy that every teacher should have as part of his or her repertoire to use when deemed most appropriate. It is hoped that what has been presented here is sufficient for the teacher to get started. Each teacher creates a cooperative learning classroom in his or her very own way. The teacher will have to make many decisions about how to proceed. Each class is different, and what works for one may not work for another. So the teacher in a cooperative learning class must be courageous, make a choice, and dive right in! Adjustments and refinements can be made as the need arises.

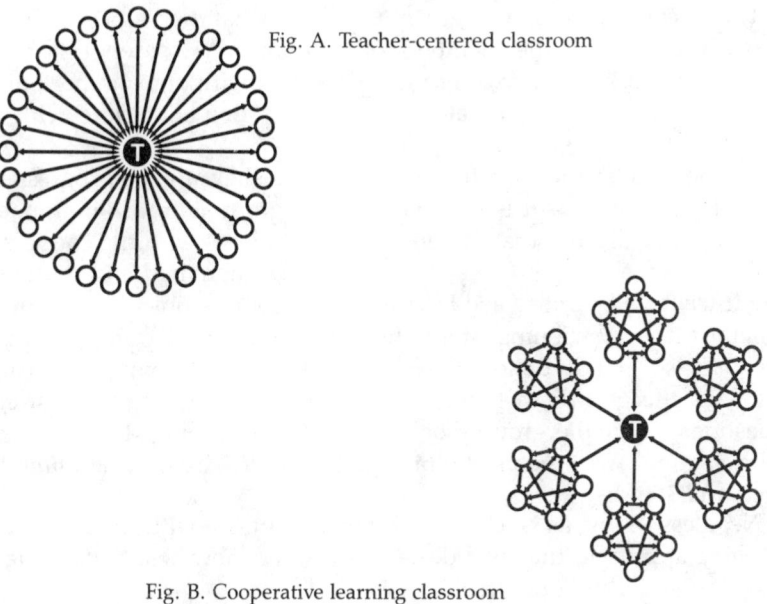

Fig. A. Teacher-centered classroom

Fig. B. Cooperative learning classroom

SAMPLE ACTIVITIES

The sample activities that follow illustrate some mathematics lessons that have been adapted for use with cooperative learning groups.

PROBLEM TOPIC 1: How Are Numbers Used in Our Lives?

Learning Level: Elementary School

Curriculum Area: Number categories; number sense

Objectives:
1. To become aware of numbers in the environment
2. To understand the different ways that numbers are used

Group Size: Three or four students

Materials for Each Group:
    Problem sheets for each student (see fig. 1)
        (or oral instructions by teacher)
    Poster paper, scissors, glue, crayons
    Old magazines and newspapers (for cutting)

Teaching Notes:
- Young children will need some help understanding what it is they are expected to do.
- Groups will need some help in seeing how numbers fall into categories according to the ways they are used: numbers for counting, identification, location, measurement, and order.
- You will want to give each group the opportunity to talk about its numbers and the way they are used. The posters will make fine additions to your bulletin board.

PROBLEM SHEET

How Are Numbers Used in Our Lives?

You will keep a Math Log for one week. Each day you will write in your log the numbers you see and the way each number is used. If possible, draw a picture, or cut out a picture from a newspaper or magazine that shows the way the number is used.

Bring in your log one week from today. You will meet with your group to discuss what you have found. The group should talk about the different ways that numbers are used. The group will then prepare a poster showing an example or a picture of each way that numbers are used.

Fig. 1. Problem sheet 1

PROBLEM TOPIC 2: What Is a Minute?

Learning Level: Elementary School

Curriculum Area: Measurement

Objectives:

1. To explore the concept of one minute
2. To experiment with the relationship between time and speed
3. To express speed as a number per minute

Group Size: Four students

Materials for Each Group:

    1 timing device
    Beads and thread
    1 ball
    4 problem sheets (fig. 2)
    1 record chart sheet (fig. 3)
    4 job cards: Timekeeper, Jumper, Bead Threader, Ball Bouncer
    1 pencil

Teaching Notes:

- If your classroom has a clock with a second hand, you may have your pupils sit quietly while the second hand sweeps around the clock for one minute.

- Young children may not be able to read the directions on their own. You need to be sure that each pupil understands what is to be done. It may be helpful to include one good reader in each group.

- Have a member of each group share the group's results with the class. Which is the activity that pupils can do the greatest number of times in one minute? Which is the activity that pupils do the smallest number of times in one minute?

- Have the groups give their ideas about things they can try to do in one minute. If possible, try some of them.

PROBLEM SHEET

What Is a Minute?

Before beginning the activity, each pupil draws a job card that tells who is the timekeeper, who is the jumper, who is the bead threader, and who is the ball bouncer. (Pupil jobs will switch in a little while.) Pupils write their names on a line of the record chart.

1. Each pupil should read aloud what he or she must do. The others should follow along silently while a pupil reads.

    Timekeeper: I am the timekeeper and recorder. I will tell the others when to start. At the end of one minute, I will tell them to stop. I will write each pupil's results in the chart.

    Jumper: When the timekeeper tells me to start, I will jump up and down and count the number of times I can jump up and down in one minute.

    Bead Threader: When the timekeeper tells me to start, I will thread beads and count the number of beads I can thread in one minute.

    Ball Bouncer: When the timekeeper tells me to start, I will bounce the ball and count the number of times I can bounce the ball in one minute.

    After the minute has ended, the timekeeper should record each pupil's count on the correct line of the record chart.

2. Pupils switch jobs with another member of the group. Each person should do something different from what he or she did before. When one minute is over, the pupil who is now timekeeper should enter each count in the record chart.

3. Switch jobs and repeat the counting activities until each pupil has done each activity once. All counts should now be in the record chart.

4. After all the results are in the record chart, talk about the results. Which is the activity that pupils can do the greatest number of times in one minute? Which is the activity that pupils do the smallest number of times in one minute?

See if your group can agree on four other things you can try to do in one minute. Write them down on the record chart sheet.

Fig. 2. Problem sheet 2

Record Chart Sheet

Record Chart

PUPIL'S NAME	Number of Jumps per Minute	Number of Beads per Minute	Number of Bounces per Minute

Other Things to Try to Do in One Minute

1. _____

_____

2. _____

_____

3. _____

_____

4. _____

_____

Fig. 3. Record sheet 2

PROBLEM TOPIC 3: Food Preferences for Students

Learning Level: Elementary School

Curriculum Area: Statistics, problem solving, fractions

Objectives:

1. To organize data in table form
2. To select appropriate data for solving problems
3. To use sample data to make predictions
4. To work with fractions using real data

Group Size: Three or four students

Materials for Each Group:

   1 problem sheet with record chart
      (each group gets a different problem sheet) (figs. 4–9)
   1 envelope containing student answers to one class survey question

Teaching Notes:

- Have each student in the class fill out and hand in a survey form (fig. 10). There should be enough questions on the form so that each group can tabulate and analyze the data for at least one question.

- When the survey forms have been collected, separate the questions. Place all student answers to question 1 in an envelope for group 1, answers to question 2 in an envelope for group 2, answers to question 3 in an envelope for group 3, and so on.

- When the groups have tabulated and analyzed the answers to the survey questions, each group will share its results and conclusions with the whole class.

PROBLEM SHEET

Food Preferences for Students (Question 1)

Your group has been given question 1 from the class's food preference survey to analyze.

Take turns!

1. One student should read the answer from each sheet. Another student should tally the results on the record sheet.
2. Total the number of tallies in each box of the table.
3. Find the totals across the bottom row. Discuss what each total means.
4. Find the totals in the last column. What does each total mean?
5. Discuss each question below and agree on an answer for each one.

   a. What fraction of the girls prefer chocolate ice cream?

   Answer _____

   b. What fraction of the students who prefer chocolate ice cream are girls?

   Answer _____

   c. What fraction of the students are girls who prefer chocolate ice cream?

   Answer _____

   d. Use the data in your table to estimate the number of boys in your grade who prefer vanilla ice cream. (Assume that there are 200 students in your grade.)

   Answer _____

Record Chart (Question 1)

	Vanilla	Chocolate	Strawberry	Total
Boys				
Girls				
Total				

Fig. 4. Problem sheet and record chart for topic 3

PROBLEM SHEET

Food Preferences for Students (Question 2)

Your group has been given question 2 from the class's food preference survey to analyze.

Take turns!

1. One student should read the answer from each sheet. Another student should tally the results on the record sheet.
2. Total the number of tallies in each box of the table.
3. Find the totals across the bottom row. Discuss what each total means.
4. Find the totals in the last column. What does each total mean?
5. Discuss each question below and agree on an answer for each one.
    a. What fraction of the girls prefer hamburgers for lunch?

    Answer _____
    b. What fraction of the students who prefer hamburgers for lunch are girls?

    Answer _____
    c. What fraction of the students are girls who prefer hamburgers for lunch?

    Answer _____
    d. Use the data in your table to estimate the number of boys in your grade who prefer frankfurters for lunch. (Assume that there are 200 students in your grade.)

    Answer _____

Record Chart (Question 2)

	Frankfurter	Hamburger	Pizza	Total
Boys				
Girls				
Total				

Fig. 5. Problem sheet and record chart for topic 3

PROBLEM SHEET

Food Preferences for Students (Question 3)

Your group has been given question 3 from the class's food preference survey to analyze.

Take turns!

1. One student should read the answer from each sheet. Another student should tally the results on the record sheet.
2. Total the number of tallies in each box of the table.
3. Find the totals across the bottom row. Discuss what each total means.
4. Find the totals in the last column. What does each total mean?
5. Discuss each question below and agree on an answer for each one.
    a. What fraction of the girls prefer carrots as a vegetable?

    Answer _____
    b. What fraction of the students who prefer carrots as a vegetable are girls?
    Answers _____
    c. What fraction of the students are girls who prefer carrots as a vegetable?

    Answer _____
    d. Use the data in your table to estimate the number of boys in your grade who prefer peas as a vegetable. (Assume that there are 200 students in your grade.)

    Answer _____

Record Chart (Question 3)

	Carrots	Cauliflower	Peas	Total
Boys				
Girls				
Total				

Fig. 6. Problem sheet and record chart for topic 3

PROBLEM SHEET

Food Preferences for Students (Question 4)

Your group has been given question 4 from the class's food preference survey to analyze.

Take turns!

1. One student should read the answer from each sheet. Another student should tally the results on the record sheet.

2. Total the number of tallies in each box of the table.

3. Find the totals across the bottom row. Discuss what each total means.

4. Find the totals in the last column. What does each total mean?

5. Discuss each question below and agree on an answer for each one.

   a. What fraction of the girls prefer soda as a drink?

      Answer _____

   b. What fraction of the students who prefer soda as a drink are girls?

      Answer _____

   c. What fraction of the students are girls who prefer soda as a drink?

      Answer _____

   d. Use the data in your table to estimate the number of boys in your grade who prefer milk as a drink. (Assume that there are 200 students in your grade.)

      Answer _____

Record Chart (Question 4)

	Apple Juice	Milk	Soda	Total
Boys				
Girls				
Total				

Fig. 7. Problem sheet and record sheet for topic 3

PROBLEM SHEET

Food Preferences for Students (Question 5)

Your group has been given question 5 from the class's food preference survey to analyze.

Take turns!

1. One student should read the answer from each sheet. Another student should tally the results on the record sheet.
2. Total the number of tallies in each box of the table.
3. Find the totals across the bottom row. Discuss what each total means.
4. Find the totals in the last column. What does each total mean?
5. Discuss each question below and agree on an answer for each one.
   a. What fraction of the girls prefer Burger King's food?

      Answer _____
   b. What fraction of the students who prefer Burger King's food are girls?

      Answer _____
   c. What fraction of the students are girls who prefer Burger King's food?

      Answer _____
   d. Use the data in your table to estimate the number of boys in your grade who prefer McDonald's food. (Assume that there are 200 students in your grade.)

      Answer _____

Record Chart (Question 5)

	Burger King	McDonald's	Wendy's	Total
Boys				
Girls				
Total				

Fig. 8. Problem sheet and record chart for topic 3

PROBLEM SHEET

Food Preferences for Students (Question 6)

Your group has been given question 6 from the class's food preference survey to analyze.

Take turns!

1. One student should read the answer from each sheet. Another student should tally the results on the record sheet.

2. Total the number of tallies in each box of the table.

3. Find the totals across the bottom row. Discuss what each total means.

4. Find the totals in the last column. What does each total mean?

5. Discuss each question below and agree on an answer for each one.

   a. What fraction of the girls prefer fruit as a snack?

   Answer _____

   b. What fraction of the students who prefer fruit as a snack are girls?

   Answer _____

   c. What fraction of the students are girls who prefer fruit as a snack?

   Answer _____

   d. Use the data in your table to estimate the number of boys in your grade who prefer pretzels as a snack. (Assume that there are 200 students in your grade.)

   Answer _____

Record Chart (Question 6)

	Fruit	Potato chips	Pretzels	Total
Boys				
Girls				
Total				

Fig. 9. Problem sheet and record chart for topic 3

SURVEY FORM

Food Preferences of Students

Directions: Answer each question by checking only one box. Be sure to indicate whether you are a boy or a girl *for each question.*

1. Which ice cream flavor do you like best?
   ☐ Chocolate   ☐ Strawberry   ☐ Vanilla
   Are you a boy or a girl? ☐ Boy   ☐ Girl

2. Which lunch do you like best?
   ☐ Frankfurter   ☐ Hamburger   ☐ Pizza
   Are you a boy or a girl? ☐ Boy   ☐ Girl

3. Which vegetable do you like best?
   ☐ Carrots   ☐ Cauliflower   ☐ Peas
   Are you a boy or a girl? ☐ Boy   ☐ Girl

4. Which drink do you like best?
   ☐ Apple juice   ☐ Milk   ☐ Soda
   Are you a boy or a girl? ☐ Boy   ☐ Girl

5. Which restaurant's food do you like best?
   ☐ Burger King   ☐ McDonald's   ☐ Wendy's
   Are you a boy or a girl? ☐ Boy   ☐ Girl

6. Which snack do you like best?
   ☐ Fruit   ☐ Potato chips   ☐ Pretzels
   Are you a boy or a girl? ☐ Boy   ☐ Girl

Fig. 10. Survey form for topic 3

PROBLEM TOPIC 4: Estimation with Money

Learning Level: Elementary School

Curriculum Area: Estimation, money, geometry

Objectives:
1. To appreciate how estimation can be used in place of computation
2. To use various methods of estimation with money
3. To learn a systematic method of recording data
4. To develop concepts of geometry

Group Size: Three students

Materials for Each Group:
   3 problem sheets (fig. 11)
   1 record chart sheet (fig. 12)
   1 calculator (to be retained by the teacher until the group hands in its
   results)

Teaching Notes:
- Appoint a reader and a recorder.
- When a group has handed in its list, you may wish to give that group a calculator so that the members can calculate the exact cost of each purchase.
- You may wish to appoint a group spokesperson to report the group's methods and results to the class.

PROBLEM SHEET

Estimation with Money

One person should read the instructions while the other group members follow along. When the reader has completed the reading, the group members should discuss what they must do.

Instructions:

1. Your group has $10 to spend. What can you buy?
   Rules: a. You must buy at least three items.
             b. You may buy more than one of an item.
             c. You may spend any amount of money—up to $10—in any way you wish.

2. Work together to list the items that your group could buy. One person should record the items on the record chart.

3. The recorder should list as many different combinations of items as the group can find. Use *estimation,* not computation, because it is faster! Show the approximate total cost of each combination of items on the record chart.

4. Hand in your group record when you are finished. KEEP A COPY OF WHAT YOU HAND IN!

Fig. 11. Problem sheet for topic 4

RECORD CHART SHEET

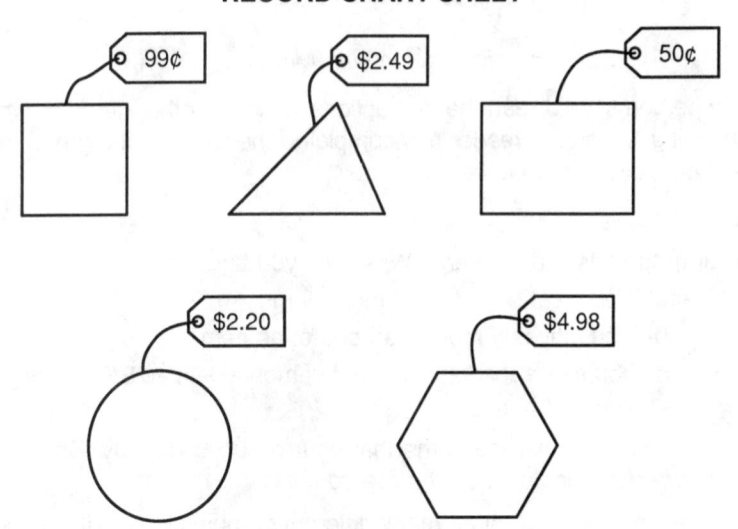

Record Chart

Purchase Number	Items Purchased	Approximate Cost of Purchase
1.		
2.		
3.		
4.		
5.		
6.		
7.		
8.		
9.		
10.		

Fig. 12. Record chart for topic 4

PROBLEM TOPIC 5: Problem Solving Using Division and the Calculator

Learning Level: Elementary and Middle School

Curriculum Area: Problem solving

Objectives:

1. To analyze problems
2. To use the calculator as a tool in problem solving
3. To practice the relationships among the dividend, the divisor, the quotient, and the remainder

Group Size: Four students

Materials for Each Group:

    4 problem sheets (fig. 13)
    1 calculator

Teaching Notes:

- Students should practice writing the relationships among the dividend, divisor, quotient, and remainder before engaging in this activity.
- Students should practice using a calculator to compute the quotient and *whole-number* remainder.

    Example: After entering $35 \div 8 =$, note that the display shows 4.375. The whole-number part of the quotient is 4. The decimal means there is a remainder. The whole-number remainder, r, is computed by using $r = 35 - (8 \times 4)$.

- In some cases, the answer to a problem will be the whole-number quotient. In other cases it will be the quotient plus 1. In some cases it will be sufficient to know that there is a remainder. In other cases it will be necessary to compute the whole-number remainder.
- The teacher may want to alert students to the fact that it is often necessary to pay sales tax on purchases or pay interest on delayed payments.
- A group gets credit for the correct solution only if any person called on in the group presents a correct explanation of the problem to the class.

PROBLEM SHEET

Problem Solving Using Division and the Calculator

All of the problems in this activity are to be solved by using a calculator to obtain information without pencil-and-paper computation. All problems involve division. However, the students in your group must analyze each problem to decide how the division results should be used to solve the problem.

TAKE TURNS USING THE CALCULATOR! One student should read the problem while the others follow along. Discuss the problem. Then proceed with the calculation and the solution.

When the members of your group have agreed on the best solution for each problem, submit *one set* of solutions for your group. Each person must be able to explain any of the group's solutions to the whole class.

1. How many quarters can you exchange for 149 pennies?
2. A class of 302 students and 5 adults are going on a field trip by bus.
   a. If each bus holds 42 persons, how many buses will be needed?
   b. Once you have determined how many buses will be needed, decide how the people should be distributed on the buses. Will additional adults be needed so that there will be one adult on each bus?
3. There are 1226 candy bars and they are to be packed 24 to a box. How many boxes will be needed? Will there be any candy bars left over?
4. A class has $10.75 to spend on notebooks and pads. Notebooks cost 69 cents and pads cost 27 cents. How many notebooks can the pupils buy if they buy as many notebooks as possible? Will there be enough money left to buy any pads? If so, how many pads can they buy?
5. A computer store advertises that a computer that costs $899 (without tax) can be purchased by paying as little as $35 a month. Mr. Day decides to buy the computer and pay $55 a month. How long will it take him to pay for the computer? How much will his last payment be?
6. Suppose today is Monday. Will it be a Monday 2646 days from today?

Fig. 13. Problem sheet for topic 5

PROBLEM TOPIC 6: Arrangements

Learning Level: Middle School or Junior High School

Curriculum Area: Permutations

Objectives:

1. To derive a formula for permutations by discerning a pattern
2. To use concrete materials as an aid to reasoning

Group Size: Four students

Materials for Each Group:

   4 problem sheets (fig. 14)
   1 record sheet for coloring squares
   5 different-colored cube blocks (same size)
   5 different-colored crayons, matching the cube colors

Teaching Notes:

- It is hoped that students will see that there are four positions for the fourth block once a three-block arrangement is in place. That is, given the following three-block arrangement,

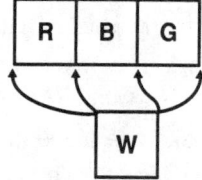

a white block can be placed as shown.

Thus, each three-block arrangement yields 4 four-block arrangements.

- There are 2 two-block arrangements, 3 · 2, or 6, three-block arrangements, and 4 · 3 · 2, or 24, four-block arrangements.
- The same line of reasoning means that a fifth block can be placed in five positions

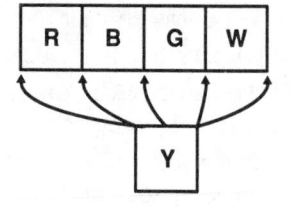

so that there are 5 · 4 · 3 · 2, or 120, different arrangements of five blocks.

PROBLEM SHEET

Arrangements

1. There are two ways to place two blocks in a row. Take turns! Place a red block and a blue block in a row. Use red and blue crayons to color the interiors of two squares to look like the blocks.

Rearrange the blocks and color the squares another way.

2. How many ways can three blocks be placed in a row?
   Use three different-colored blocks. Take turns!

   a. Arrange the three blocks in a row.
   b. Color squares on the record sheet to show each arrangement.
   c. How many arrangements are there?

3. How many ways can four blocks be arranged in a row?
   Make four-block arrangements from each three-block arrangement. Share the work! Work in pairs! Take turns!

   a. Make *one* of the three-block arrangements.
      (1) Now take a fourth block of a different color and make a four-block arrangement from the three-block arrangement.
      (2) Color squares to show this four-block arrangement.
      (3) Move the fourth block to make a different four-block arrangement. Color squares to show this arrangement.
      (4) Make as many four-block arrangements as you can from the same three-block arrangement. Keep a record by coloring squares on your record sheet.
      (5) How many four-block arrangements have you made?
   b. Make four-block arrangements from each three-block arrangement. Keep a record by coloring squares.
   c. Discuss your findings with your teammates. Is there a pattern? How many four-block arrangements are there altogether?

4. How many five-block arrangements are there?
   First you must agree on an easy way to solve this problem. Then answer the question.

Fig. 14. Problem sheet for topic 6

RECORD CHART SHEET

Three-block arrangements

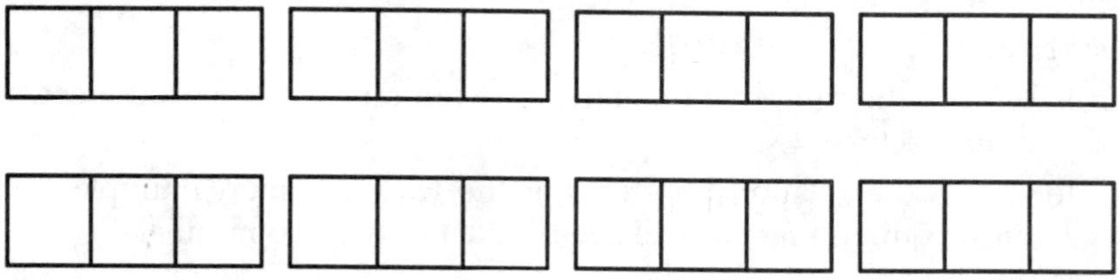

Four-block arrangements

Fig. 15. Record chart for topic 6

PROBLEM TOPIC 7: Least Common Multiple and
Greatest Common Factor

Learning Level: Junior High School

Curriculum Area: Number theory

Objectives:

1. To practice the computation of the least common multiple and greatest common factor
2. To derive the relationships among the least common multiple, the greatest common factor, and the product of the two numbers
3. To record data systematically
4. To discern a pattern by analyzing data

Group Size: Four students

Materials for Each Group:

 4 copies of problem sheet (fig. 16)
 1 record sheet (fig. 17)
 1 number envelope containing 12 slips of paper, each with a pair of numbers

Teaching Notes:

- Appoint a reader and a recorder for each group.
- If you decide to have a group report its findings to the entire class, choose a spokesperson for the group after all record sheets have been submitted.
- These are pairs of numbers (m, n) that are relatively prime; that is, 1 is the greatest common factor:
  (1,3), (2,3), (3,5), (4,7), (6,11), (8,15), (6,35), (13,17), (3,46), (15,26), (21,5)
- These pairs of numbers (m,n) have a common factor greater than 1:
  (3,6), (6,8), (8,12), (12,26), (12,15), (10,12), (30,45), (3,645), (15,65)

PROBLEM SHEET

Least Common Multiple and Greatest Common Fa

One person should read the instructions while other group members follow along. When the reader has completed the reading, others may question or explain the tasks and the requirements of the problem. When the group is ready to begin, each member should select, at random, three slips of paper from the number envelope.

1. Each person will receive three pairs of numbers. Analyze each pair of numbers, (m,n), to determine—
   a. the greatest common factor of m and n: $GCF(m,n)$;
   b. the least common multiple of m and n: $LCM(m,n)$;
   c. the product of m and n: $m \cdot n$.
   Example: $GCF(4,6) = 2$, $LCM(4,6) = 12$, $m \cdot n = 24$.

2. Each group member who has completed his or her part of the work should offer to help another member.

3. Students who have completed their work should exchange papers and check one another's results.

4. When all pairs of numbers have been analyzed, the recorder should list all group results on the record sheet.

5. When the group has agreed on the results, the members should discuss their findings and determine the relationships among the greatest common factor, the least common multiple, and the product of any two numbers.

6. State the relationship. Then test it out on four new pairs of numbers, one pair chosen by each group member.

7. The recorder should record the group's work in #6 on the record sheet. When the group has agreed on what is written on the record sheet, hand it in.

Fig. 16. Problem sheet for topic 7

RECORD SHEET

Number Pairs (m,n)	GCF (m,n)	LCM (m,n)	Product (m · n)

Relationship:

Fig. 17. Record sheet for topic 7

PROBLEM TOPIC 8: Properties of Parallelograms

Learning Level: Junior High School or Secondary School

Curriculum Area: Geometry

Objectives:

1. To review the definitions of parallelogram, rectangle, rhombus, and square
2. To practice using a protractor to measure angles
3. To discern a pattern among the opposite angles of a parallelogram
4. To discern a pattern among the opposite sides of a parallelogram

Group Size: Four students

Materials for Each Group:

    4 problem sheets (fig. 18)
    1 set of parallelogram sheets (figs. 19–22)
    1 summary record sheet (fig. 23)
    4 rulers
    4 protractors

Teaching Notes:

- One student in each group should read the Problem Sheet to the group while the others follow along.
- One student in each group should record the measurements and the *relationship statements* on the summary record sheet.
- If you wish to have a group report to the whole class, make a random selection of a student who will be the spokesperson for the group.
- A follow-up activity might involve the diagonals of the parallelogram.

PROBLEM SHEET

Properties of Parallelograms

There are four different parallelograms. Your group is to measure the sides and the angles of each one.

1. Split up the work! Each student should start with one parallelogram. Measure one side and one angle of the parallelogram. Be sure to record your measurements directly on the parallelogram sheet in the space provided.

2. Pass your sheet to another student in your group who will measure *one* side and *one* angle and record the measurements on the sheet.

3. Continue exchanging parallelograms until all measurements have been made.

4. Once all parallelograms have been measured, take turns dictating the information to the recorder, who will record it on the summary record sheet. If a particular measurement is challenged, the measurement should be made again in the presence of the other group members.

5. Once the group members agree on the data that has been entered on the summary record sheet, examine the data to see whether any patterns appear. Is there a relationship between particular sides of all parallelograms? Is there a relationship between particular angles of all parallelograms?

6. When you have completed the summary record sheet, hand it in.

Fig. 18. Problem sheet for topic 8

PARALLELOGRAM SHEET

I

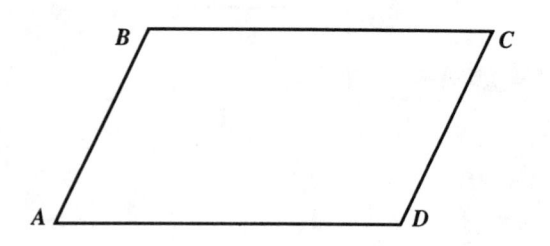

$AB =$ _____ $BC\ \ =$ _____

$CD =$ _____ $DA\ \ =$ _____

$m\angle A =$ _____ $m\angle B =$ _____

$m\angle C =$ _____ $m\angle D =$ _____

Fig. 19. Parallelogram sheet for topic 8

PARALLELOGRAM SHEET

II

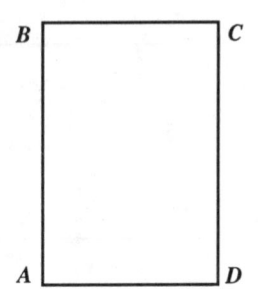

$AB =$ _____ $BC \;=$ _____

$CD =$ _____ $DA \;=$ _____

$m\angle A =$ _____ $m\angle B =$ _____

$m\angle C =$ _____ $m\angle D =$ _____

Fig. 20. Parallelogram sheet for topic 8

PARALLELOGRAM SHEET

III

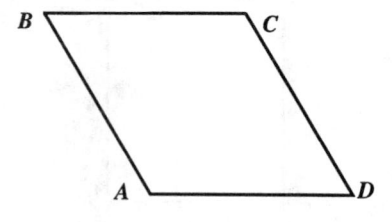

$AB =$ _____ $BC \ =$ _____

$CD =$ _____ $DA \ =$ _____

$m\angle A =$ _____ $m\angle B =$ _____

$m\angle C =$ _____ $m\angle D =$ _____

Fig. 21. Parallelogram sheet for topic 8

PARALLELOGRAM SHEET

IV

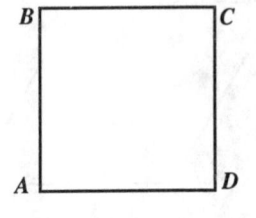

AB = _____   BC = _____

CD = _____   DA = _____

$m\angle A$ = _____ $\angle B$ = _____

$m\angle C$ = _____ $m\angle D$ = _____

Fig. 22. Parallelogram sheet for topic 8

SUMMARY RECORD SHEET

Measurements

Parallelogram	AB	CD	BC	DA	$m\angle A$	$m\angle B$	$m\angle C$	$m\angle D$
I								
II								
III								
IV								

Relationship (sides) _____

Relationship (angles) _____

Fig. 23. Summary record sheet for topic 8

PROBLEM TOPIC 9: Coin-Die Experiment

Learning Level: Junior High School or Secondary School

Curriculum Area: Probability

Objectives:
1. To conduct a probability experiment and record the results
2. To use the relative frequency definition of probability
3. To determine the mathematical probability of a compound event
4. To compare the relative frequency and mathematical probability of an event

Group Size: Three students

Materials for Each Group:
   3 problem sheets (fig. 24)
   1 record sheet (fig. 25)
   1 coin
   1 die

Teaching Notes:
- The 12 possible outcomes of the experiment appear on the record sheet. However, you may wish to discuss the fact that this number can be computed by using the multiplication principle: that is, 2×6.
- If your students are more advanced, you may wish them to construct their own record sheet.

PROBLEM SHEET

Coin-Die Experiment

One student should read through the instructions while the others follow along. Discuss the problem before proceeding.

For this experiment a trial is a toss of the coin and a roll of the die. Your group is to record the results of 50 trials on the accompanying record sheet.

1. One student should roll the die while a second student tosses the coin. The third student should tally the results on the record sheet.

2. After 25 trials, change roles.

3. Now that you have completed the experiment, answer the following questions:

   a) Experimental Probability

   (1) Use the relative frequency definition of probability to determine the probability of both a 5 and tails occurring.

   $P(5,T) = $ _____

   (2) If your group were to conduct another 100 trials, approximately how many times would you expect to see a 5 and tails?

   _____

   b) Mathematical Probability

   (1) How many outcomes are there for the compound event "roll a die and toss a coin"?

   _____

   (2) Are the outcomes equally likely?

   _____

   (3) What is the mathematical probability of both a 5 and tails occurring?

   $P(5,T) = $ _____

   c) Compare your results in parts *a* and *b* above. How could you change the experiment so that the result in part *a* is closer to the mathematical probability obtained in part *b*?

   _____

Fig. 24. Problem sheet for topic 9

RECORD SHEET

Outcome	Tally	Frequency	Outcome	Tally	Frequency
(1,Heads)			(1,Tails)		
(2,Heads)			(2,Tails)		
(3,Heads)			(3,Tails)		
(4,Heads)			(4,Tails)		
(5,Heads)			(5,Tails)		
(6,Heads)			(6,Tails)		

Fig. 25. Record chart sheet for topic 9

PROBLEM TOPIC 10: Using Parentheses

Learning Level: Secondary School

Curriculum Area: Algebra

Objectives:

1. To use parentheses in the computation of numerical expressions.
2. To use the "working backwards" strategy to solve a problem

Group Size: Three or four students

Materials for Each Group:

   1 problem sheet (fig. 26)

Teaching Notes:

- Students may try to solve these problems by using a trial-and-error placement of parentheses. The teacher should make the students aware of the advantage of using a working backwards strategy.
- The teacher should randomly select a member of each group to explain the group's solutions.

PROBLEM SHEET

Using Parentheses

The value of an expression depends upon how parentheses are used.

1. Work together to place parentheses, when needed, to complete true statements.

$$2 + 3 \cdot 5^2 = 77$$
$$2 + 3 \cdot 5^2 = 125$$
$$2 + 3 \cdot 5^2 = 227$$
$$2 + 3 \cdot 5^2 = 289$$

2. Make up another set of problems similar to the ones given. Your teacher will give them to another group to solve.

3. When you have finished, hand in your group's work.

Fig. 26. Problem sheet for topic 10

PROBLEM TOPIC 11: Probability with Linear Equations

Learning Level: Secondary School

Curriculum Area: Algebra, combinatorics, and probability

Objectives:

1. To apply permutations in the assignment of coefficients
2. To practice solving linear equations
3. To compute probabilities from a list of outcomes

Group Size: Three students

Materials for Each Group:

   1 problem sheet with record chart (figs. 27 and 28)

Teaching Notes:

- The teacher should randomly select one person from a group to explain the group's solution.
- The approach outlined for this topic can also be used in the activity, "Probability with Quadratic Equations."

PROBLEM SHEET

Probability with Linear Equations

The numbers 2, 3, and 5 are substituted at random for p, q, and r in the equation $px + q = r$ $(p \neq q \neq r)$.

- What is the probability that the solution is negative?
- What is the probability that the absolute value of the solution is 1?
- What is the probability that the solution is a fraction?
- If r is not 5, what is the probability that the solution is negative?

1. Engage in a group discussion of the problem and possible strategies for solution.

2. Work as a group to write down all the different possible substitutions of the numbers 2, 3, and 5 for p, q, and r. Use the record chart to record your results.

3. Complete the record chart by solving each equation. Be sure to divide the work so that each person in the group solves the same number of equations.

4. Compute each probability:

   a) Probability that the solution is negative

   _____

   b) Probability that the absolute value of the solution is 1

   _____

   c) Probability that the solution is a fraction

   _____

   d) Probability that the solution is negative when r is not 5

   _____

Fig. 27. Problem sheet for topic 11

RECORD CHART

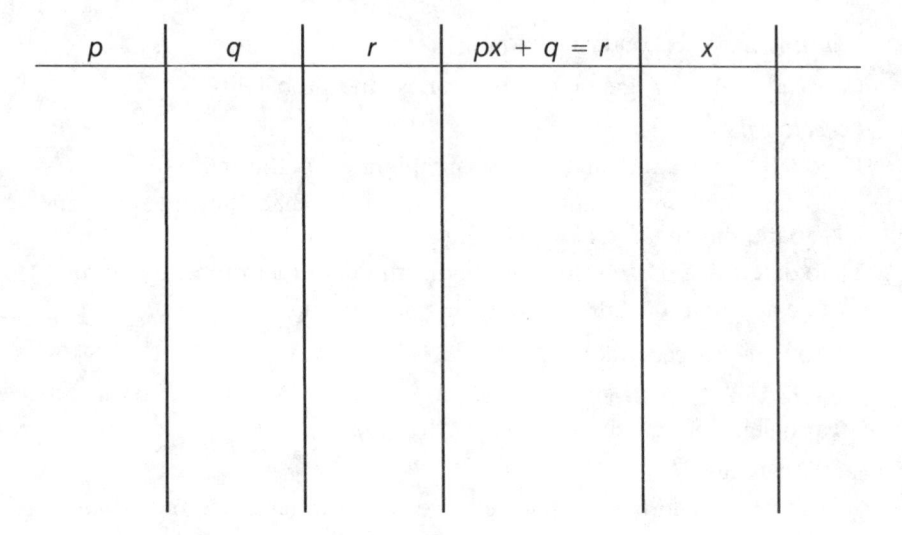

p	q	r	$px + q = r$	x	

Fig. 28. Record chart for topic 11

PROBLEM TOPIC 12: Probability with Quadratic Equations

Learning Level: Secondary School

Curriculum Area: Algebra, combinatorics, and probability

Objectives:

1. To apply permutations in the assignment of coefficients
2. To determine those values of the discriminant that allow the quadratic equation to be solved by factoring
3. To determine those values of the discriminant that produce real roots
4. To compute probabilities from a list of outcomes

Group Size: Three students

Materials for Each Group:

   1 problem sheet with record chart (fig. 29)

Teaching Notes:

- Students should find that since there are six different permutations of a, b, and c, there are six different equations to consider.

PROBLEM SHEET

Probability with Quadratic Equations

The numbers 1, 3, and 4 are substituted at random for a, b, and c in the quadratic equation $ax^2 + bx + c = 0$ ($a \neq b \neq c$).

- What is the probability that $ax^2 + bx + c = 0$ can be solved by factoring?
- What is the probability that $ax^2 + bx + c = 0$ has real roots?

1. Engage in a group discussion of the problem and possible strategies for solution.

2. Work as a group to write down all the different possible substitutions of the numbers 1, 3, and 4 for a, b, and c. Use the record chart to record your results.

3. Complete the record chart by computing the discriminant for each equation.

4. Compute each probability.

   a) Probability that the equation can be solved by factoring

   _____

   b) Probability that the equation has real roots

   _____

RECORD CHART

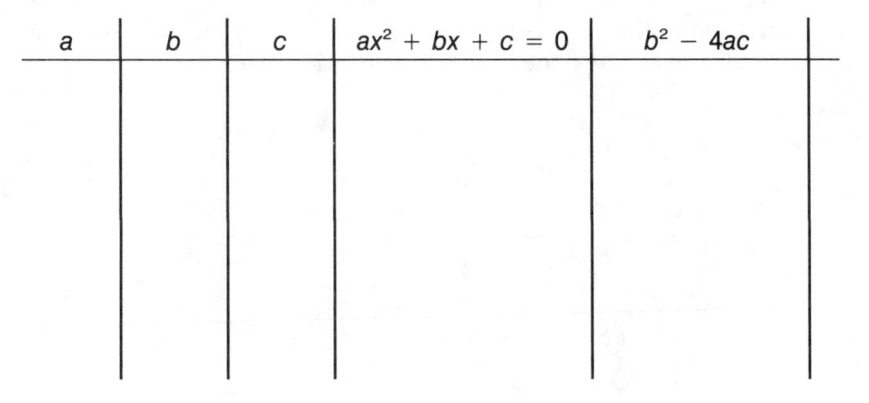

a	b	c	$ax^2 + bx + c = 0$	$b^2 - 4ac$	

Fig. 29. Problem sheet and record chart for topic 12

PROBLEM TOPIC 13: Using Transformations to Graph Parabolas

Learning Level: Secondary School

Curriculum Area: Parabolas, transformation, geometry

Objectives:

1. To graph the parabola $y = x^2 + k$ (k a real number) by translating the graph of the parabola $y = x^2$.

2. To graph the parabola $y = (x-h)^2$ (h a real number) by translating the graph of the parabola $y = x^2$.

3. To graph the parabola $y = (x-h)^2 + k$ (h, k real numbers) by using a composition of translations of the parabola $y = x^2$.

Group Size: Three students

Materials for Each Group:

  1 problem sheet (fig. 30)
  Graph paper
  Tracing paper

Teaching Notes:

- Activity 1: To superimpose the graph of $y = x^2$ on the graphs of the form $y = x^2 + k$ (k a real number), the students must *translate* the graph $y = x^2$, k units in the direction of the y axis.

- Activity 2: To superimpose the graph of $y = x^2$ on the graphs of the form $y = (x-h)^2$ (h a real number), the students must *translate* the graph $y = x^2$, h units in the direction of the x axis.

- Activity 3: As a result of activities 1 and 2 the students will see that to obtain the graph of $y = (x-h)^2 + k$ (h, k are real numbers) from the graph $y = x^2$, they must *translate* the graph $y = x^2$, k units in the direction of the y axis and h units in the direction of the x axis. The translation can be done in the reverse order as well.

PROBLEM SHEET

Using Transformations to Graph Parabolas

Activity 1

1. On separate pieces of graph paper sketch the graph of each of the following equations:

   Student A: $y = x^2$
   Student B: $y = x^2 + 3$
   Student C: $y = x^2 - 4$

2. Student A: Use the tracing paper to trace your graph of $y = x^2$. (Trace the axes as well.)

3. In the group, use the tracing to discern a relationship among the three parabolas. Agree on the relationship, and write it in the space provided.

   Relationship: _____

   a) Place the tracing of the graph of $y = x^2$ on the graph of $y = x^2 + 3$ so that the *axes* coincide.
      Work together to state the transformation that can be used to obtain the graph of $y = x^2 + 3$ from the graph of $y = x^2$.

   Transformation: _____

   b) Place the tracing of the graph of $y = x^2$ on the graph of $y = x^2 - 4$ so that the *axes* coincide.
      State the transformation that can be used to obtain the graph of $y = x^2 - 4$ from the graph of $y = x^2$.

   Transformation: _____

4. Agree on a general method for obtaining the graph of $y = x^2 + k$ (k is a real number) from the graph of $y = x^2$.

   Generalization: _____

5. a) All students: Draw the graph of $y = x^2$.
   b) Use the generalization you have agreed on to draw the following graphs:

   Student A: $y = x^2 - 2$
   Student B: $y = x^2 - 4.5$
   Student C: $y = x^2 + 5$

   Check each other's graphs.

Activity 2

1. Draw the graphs of each of the following congruent parabolas:
$$\text{Student A: } y = (x - 1)^2$$
$$\text{Student B: } y = (x + 2)^2$$
$$\text{Student C: } y = (x - 3)^2$$

2. Superimpose the graph of $y = x^2$ on each of the three graphs in (1) so that, in each case, the *axes* coincide.

   In each case the group should state the transformation that can be used to obtain the new graph from the graph of $y = x^2$.

   Transformations:

   $y = (x - 1)^2$: _____

   $y = (x + 2)^2$: _____

   $y = (x - 3)^2$: _____

3. Find a generalization for a method of obtaining the graph of $y = (x - h)^2$ (h is a real number) from the graph of $y = x^2$.

   Generalization: _____

4. Use the generalization you have agreed on to graph each of the following parabolas:
$$\text{Student A: } y = (x + 3)^2$$
$$\text{Student B: } y = (x - 4)^2$$
$$\text{Student C: } y = (x + 1)^2$$

   Check each other's graphs.

Activity 3

1. Decide within the group how one can best use the discoveries made in the two previous activities to draw the graph of $y = (x - 1)^2 + 3$. Draw the graph.

2. State a generalization for a method of obtaining the graph of $y = (x - h)^2 + k$ (h, k are real numbers) from the graph of $y = x^2$.
   Generalization: _____

3. Draw the graph of $y = x^2 - 2x + 4$. Compare this graph with the graph of $y = (x - 1)^2 + 3$. What do you notice? Why is this so?

   _____

   _____

Activity 4

1. Prepare a group summary of what you have learned from these activities.

2. Suggest other similar explorations that might be made with the graphs of parabolas.

Fig. 30. Problem sheet for topic 13

PROBLEM TOPIC 14: Review of Circles

Learning Level: Secondary School

Curriculum Area: Geometry

Objective:

To review the application of circle theorems

Group Size: Four or five students

Materials for Each Group:

1 problem sheet for each student (fig. 31)

Teaching Notes:

- The students are to complete the problem sheet as a homework assignment. The next day of class the students must agree on the solutions to the problems. A group gets credit for the correct solution only if any person who is called on in the group presents a correct explanation of the problem to the class.
- Similar problem sheets can be made to review work in other areas of the curriculum.

PROBLEM SHEET

Review of Circles

1. Complete the given exercises at home. State all theorems you used in each case.

2. Bring your completed assignment to class tomorrow for group discussion.

3. When the members of your group have agreed on the best solution for each problem, submit *one set* of solutions for your group.

4. Each person must be able to explain any of the group's solutions to the whole class.

Exercises

a) In a circle whose radius is 10, a chord is 16 units long. Compute the distance of the chord from the center of the circle.

b) Two concentric circles have radii of length 5 and 13 centimeters. Compute the length of a chord of the larger circle that is tangent to the smaller circle.

c) From a point, P, outside circle O, a tangent $\overline{PC}$ is drawn. If the radius of the circle measures 8 centimeters and the tangent measures 15 centimeters, what is the distance, PO, from the point P to the center of the circle?

d) In circle O, diameter $\overline{AB}$ measures 14 units. If chord $\overline{AC}$ measures 7 units, calculate the measure of angle CAB.

e) The segment joining the midpoint of a chord to the midpoint of its minor arc has length 4 cm. If the chord itself has length 20 cm, find the length of the diameter of the circle.

f) An equilateral pentagon is inscribed in a circle. Compute the measure of the angle formed by a side of the pentagon and a line tangent to the circle at one of the vertices of the pentagon.

Fig. 31. Problem sheet for topic 14

PROBLEM TOPIC 15: Roots and Coefficients of Quadratic Equations

Learning Level: Secondary School

Curriculum Area: Algebra

Objectives:
1. To practice solving quadratic equations by factoring
2. To determine the relationship between the coefficients and the sum of the roots
3. To determine the relationship between the coefficients and the product of the roots

Group Size: Four students

Materials for Each Group:
 4 problem sheets (fig. 32)
 1 record sheet (fig. 33)
 3 problem envelopes:
   Envelope 1 contains four slips of paper, each with a different quadratic equation of the form $ax^2 + bx + c = 0$ $(a = 1)$.
   Envelope 2 contains four slips of paper, each with a different quadratic equation of the form $ax^2 + bx + c = 0$ $(a \neq 0$ and $a \neq 1)$.
   Envelope 3 contains four slips of paper, each with a quadratic equation of the form $ax^2 + bx + c = 0$ $(a \neq 0)$ different from those in envelopes 1 and 2.

Teaching Notes:
- In problem envelope 1, include four slips of paper. On each paper write a different equation (suggestions are listed below).
  1) $x^2 + 2x - 15 = 0$ 2) $x^2 + x - 6 = 0$
  3) $x^2 - 9x + 20 = 0$ 4) $x^2 - 4x - 12 = 0$
  Since $a = 1$ in each of the equations above, the relationship is $r_1 + r_2 = -b$ and $r_1 \cdot r_2 = c$.
- In problem envelope 2, include four slips of paper. On each paper write a different equation (suggestions are listed below).
  1) $2x^2 + x - 3 = 0$ 2) $3x^2 + 11x - 4 = 0$
  3) $20x^2 + 19x + 3 = 0$ 4) $6x^2 - 19x + 10 = 0$
  In these equations the students will notice that $r_1 + r_2 = -b/a$ and $r_1 \cdot r_2 = c/a$. They will then check to see that this relationship holds for the cases when $a = 1$.

PROBLEM SHEET

Roots and Coefficients of Quadratic Equations

Activity 1

One person should read the instructions while other group members follow along. When the reader has completed the reading, others may question or explain the tasks and the requirements of the problem. When the group is ready to begin, each member should select, at random, one slip of paper from problem envelope 1.

1. Each person will receive a quadratic equation of the form $ax^2 + bx + c = 0$ $(a = 1)$, that has roots r_1 and r_2. Analyze each equation to determine—

   a) the values of a, b, and c;

   b) the roots r_1 and r_2;

   c) the sum of the roots: $r_1 + r_2$;

   d) the product of the roots: $r_1 \cdot r_2$.

2. Each group member who has completed his or her part of the work should offer to help another member.

3. Students who have completed their work should exchange papers and check one another's results.

4. When all the equations have been analyzed and the group has agreed on the results, one student should list the results in the record sheet.

5. The group members should discuss their findings and determine—

   a) the relationship among the coefficients a, b, and c and the sum of the roots;

   _____

   b) the relationship among the coefficients a, b, and c and the product of the roots.

   _____

Activity 2

1. Each member of the group should select, at random, one slip of paper from problem envelope 2. Each person will receive a quadratic equation of the form $ax^2 + bx + c = 0$ $(a \neq 0, a \neq 1)$, with the roots r_1 and r_2. Analyze each equation as in activity 1. (Find the values of a, b, c, r_1, r_2, $r_1 + r_2$, and $r_1 \cdot r_2$.)

2. When all of the equations have been analyzed and the group has agreed on the results, one student should list the results on the record sheet.

3. The group members should discuss their findings and determine the relationship among the coefficients *a, b,* and *c* of the equation, and the sum and product of the roots.

4. State the relationships.

_____

_____

5. Check to see that these relationships among the roots and coefficients apply to the previous equations having $a = 1$.

Activity 3

1. Each member of the group should select, at random, one slip of paper from problem envelope 3. Each person will receive a quadratic equation in the form $ax^2 + bx + c = 0$. *Without solving the equation,* determine the sum and product of the roots. Write your answers on the paper.

2. Exchange papers. *Solve* the equation on the new slip of paper. Check to see that the sum and product of the roots, written on the slip of paper, agree with your solution.

3. If the results *do not agree,* the two students who determined the roots and the sum and product of the roots should discuss their findings and come to an agreement.

Fig. 32. Problem sheet for topic 15

RECORD SHEET

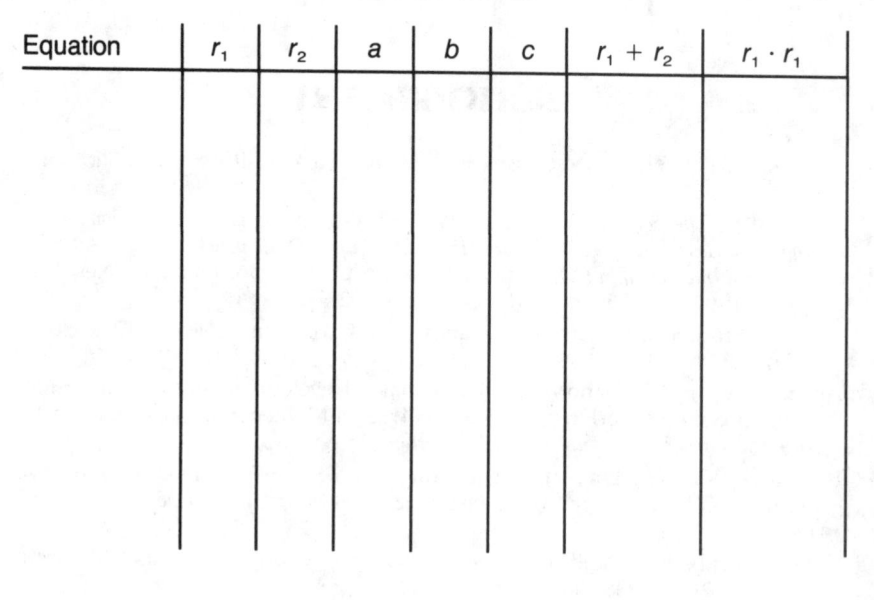

Equation	r_1	r_2	a	b	c	$r_1 + r_2$	$r_1 \cdot r_1$

Fig. 33. Record sheet for topic 15

BIBLIOGRAPHY

Aronson, Elliot. *The Jigsaw Classroom*. Newbury Park, Calif.: Sage Publications, 1978.

Artzt, Alice. *The Comparative Effects of the Student-Team Method of Instruction and the Traditional Teacher-centered Method of Instruction upon Student Achievement, Attitude, and Social Interaction in High School Mathematics Courses*. Doctoral diss., New York University, 1983. University Microfilms (84-06277).

———. "Student Teams in Mathematics Class." *Mathematics Teacher* 72 (October 1979): 505–8.

Artzt, Alice F., and Eleanor Armour-Thomas. "Development of a Cognitive-Metacognitive Framework for Protocol Analysis of Mathematical Problem Solving in Small Groups." *Cognition and Instruction*, in review.

Behounek, Karla J., Linda J. Rosenbaum, Les Brown, and Janet V. Burcalow. "Our Class Has Twenty-five Teachers." *Arithmetic Teacher* 36 (December 1988): 10–13.

Cohen, Elizabeth G., and Joan Benton. "Making Groupwork Work." *American Educator* 12 (Fall 1988): 10–17, 45–46.

Davidson, Neil. "Small-Group Cooperative Learning in Mathematics: A Review of the Research." In *Cooperative Learning Research in Mathematics*, edited by Neil Davidson and Roberta Dees. Proposed monograph of the *Journal for Research in Mathematics Education*.

———. "Small-Group Learning and Teaching in Mathematics: A Selective Review of the Research." In *Learning to Cooperate, Cooperating to Learn*, edited by Robert Slavin, Shlomo Sharan, Spencer Kagan, Rachel Hertz-Lazarowitz, Clark Webb, and Richard Schmuck. New York: Plenum Publishing Corp., 1985.

Davidson, Neil, ed. *Cooperative Learning in Mathematics: A Handbook for Teachers*. Menlo Park, Calif.: Addison-Wesley, 1990.

Dishon, Dee, and Pat O'Leary. *A Guidebook for Cooperative Learning: A Technique for Creating More Effective Schools*. Holmes Beach, Fla.: Learning Publications, 1984.

Gilbert-Macmillan, Kathleen, and Steven Leitz. "Cooperative Small Groups: A Method for Teaching Problem Solving." *Arithmetic Teacher* 33 (March 1986): 9–11.

Johnson, David W., and Roger T. Johnson. *Cooperation and Competition: Theory and Research*. Edina, Minn.: Interaction Book Co., 1989.

———. "Critical Thinking through Structured Controversy." *Educational Leadership* 45 (1988): 58–64.

———. "Instructional Goal Structure: Cooperative, Competitive, or Individualistic." *Review of Educational Research* 44 (1974): 213–40.

———. *Learning Together and Alone: Cooperative, Competitive, and Individualistic Learning*. 2d ed. Englewood Cliffs, N.J.: Prentice Hall, 1987.

Johnson, David W., Roger T. Johnson, and Edythe Johnson Holubec. *Revised Circles of Learning: Cooperation in the Classroom*. Edina, Minn.: Interaction Book Co., 1986.

Kagan, Spencer. *Cooperative Learning Resources for Teachers*. 4th ed. Laguna Niguel, Calif.: Resources for Teachers, 1987.

———. "The Structural Approach to Cooperative Learning." *Educational Leadership* 47 (December 1989/January 1990): 12–15.

National Council of Teachers of Mathematics. *Curriculum and Evaluation Standards for School Mathematics*. Reston, Va.: The Council, 1989.

Rosenbaum, Linda, Karla J. Behounek, Les Brown, and Janet V. Burcalow. "Step into Problem Solving with Cooperative Learning." *Arithmetic Teacher* 36 (March 1989): 7–11.

Sharan, Shlomo. "Cooperative Learning in Small Groups: Recent Methods and Effects on Achievement, Attitudes and Ethnic Relations." *Review of Educational Research* 50 (1980): 241–71.

Sharan, Yael, and Shlomo Sharan. "Group Investigation Expands Cooperative Learning." *Educational Leadership* 47 (December 1989/January 1990): 17–21.

Slavin, Robert E. "Cooperative Learning." *Review of Educational Research* 50 (1980): 315–42.

———. *Cooperative Learning*. New York: Longman, 1983.

———. "Cooperative Learning and Individualized Instruction." *Arithmetic Teacher* 35 (November 1987): 7–13.

———. "Cooperative Learning and Student Achievement." *Educational Leadership* 46 (October 1988): 31–33.

———. *Using Student Team Learning*. Rev. ed. Baltimore: Center for Social Organization of Schools, Johns Hopkins University, 1980.

———. "When Does Cooperative Learning Increase Student Achievement?" *Psychological Bulletin* 94 (1983): 429–45.

Slavin, Robert E., Marshall Leavey, and Nancy Madden. "Combining Cooperative Learning and Individualized Instruction: Effects on Student Mathematics Achievement, Attitudes, and Behaviors." *Elementary School Journal* 84 (1984): 7–13.

Slavin, Robert E., Nancy Madden, and Marshall Leavey. "Effects of Team Assisted Individualization on the Mathematics Achievement of Academically Handicapped and Nonhandicapped Students." *Journal of Educational Psychology* 76 (1984): 813–19.

Slavin, Robert E., Shlomo Sharan, Spencer Kagan, Rachel Hertz-Lazarowitz, Clark Webb, and Richard Schmuck. *Learning to Cooperate, Cooperating to Learn*. New York: Plenum Publishing Corp., 1985.

Suydam, Marilyn. "Research Report: Individualized or Cooperative Learning." *Arithmetic Teacher* 32 (April 1985): 39.

Webb, Noreen M. "Student Interaction and Learning in Small Groups: A Research Summary." In *Learning to Cooperate, Cooperating to Learn*, edited by Robert Slavin, Shlomo Sharan, Spencer Kagan, Rachel Hertz-Lazarowitz, Clark Webb, and Richard Schmuck. New York: Plenum Publishing Corp., 1985.

———. "Student Interaction and Mathematics Learning in Small Groups." In *Cooperative Learning Research in Mathematics*, edited by Neil Davidson and Roberta Dees. Proposed monograph of the *Journal for Research in Mathematics Education*.